ANIMAL
Amphibians

by Kari Schuetz

MW01610483

BLASTOFF! READERS 3

BELLWETHER MEDIA • MINNEAPOLIS, MN

Note to Librarians, Teachers, and Parents:

Blastoff! Readers are carefully developed by literacy experts and combine standards-based content with developmentally appropriate text.

Level 1 provides the most support through repetition of high-frequency words, light text, predictable sentence patterns, and strong visual support.

Level 2 offers early readers a bit more challenge through varied simple sentences, increased text load, and less repetition of high-frequency words.

Level 3 advances early-fluent readers toward fluency through increased text and concept load, less reliance on visuals, longer sentences, and more literary language.

Level 4 builds reading stamina by providing more text per page, increased use of punctuation, greater variation in sentence patterns, and increasingly challenging vocabulary.

Level 5 encourages children to move from "learning to read" to "reading to learn" by providing even more text, varied writing styles, and less familiar topics.

Whichever book is right for your reader, Blastoff! Readers are the perfect books to build confidence and encourage a love of reading that will last a lifetime!

This edition first published in 2016 by Bellwether Media, Inc.

No part of this publication may be reproduced in whole or in part without written permission of the publisher. For information regarding permission, write to Bellwether Media, Inc., Attention: Permissions Department, 6012 Blue Circle Dr., Minnetonka, MN 55343.

Library of Congress Cataloging-in-Publication Data
Schuetz, Kari.
 Amphibians / by Kari Schuetz.
    p. cm. – (Blastoff! readers: animal classes)
 Includes bibliographical references and index.
 Summary: "Simple text and full-color photography introduce beginning readers to amphibians. Developed by literacy experts for students in kindergarten through third grade"–Provided by publisher.
 ISBN: 978-1-60014-771-5 (hardcover : alk. paper)
 ISBN: 978-1-62617-490-0 (paperback : alk. paper)
 1. Amphibians–Juvenile literature. I. Title.
 QL644.2.S3365 2013
 597.8–dc23   2012000963

Text copyright © 2013 by Bellwether Media, Inc. BLASTOFF! READERS and associated logos are trademarks and/or registered trademarks of Bellwether Media, Inc.

Printed in the United States of America, North Mankato, MN.

Table of Contents

Alive or **extinct**, every animal is part of the animal kingdom.

They are sorted into groups based on common features.

What Are Amphibians?

The amphibian **class** includes animals that live part of their life in water and part on land. They are **vertebrates** because they have backbones.

The Animal Kingdom

vertebrates

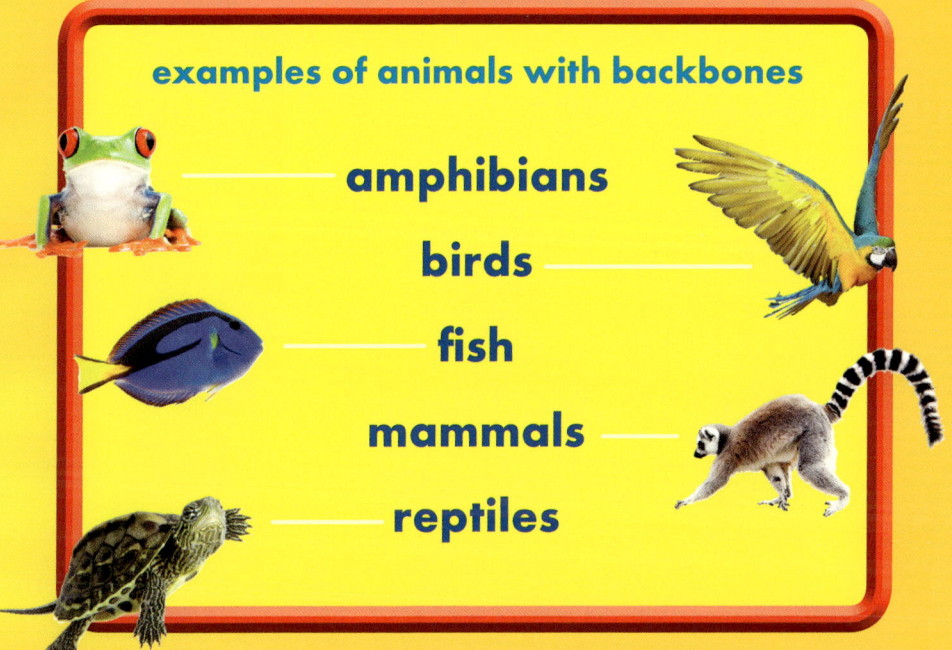

examples of animals with backbones

amphibians

birds

fish

mammals

reptiles

invertebrates

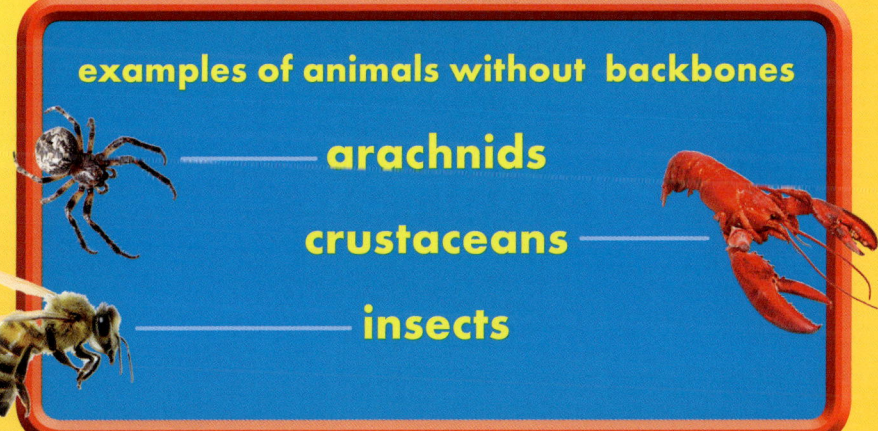

examples of animals without backbones

arachnids

crustaceans

insects

Amphibians are **cold-blooded** animals. Their bodies are the same temperature as their surroundings.

burrow

Too much heat can kill amphibians. They spend a lot of time in wet, shaded areas and in **burrows**.

Water and **oxygen** can pass through the skin of most amphibians. Many have a layer of slime to keep their skin wet.

Amphibians **shed** their skin as they grow. Some eat their dead skin for its **nutrients**.

Most amphibians hatch from eggs that are laid in water. Many have **gills** to breathe underwater and tails to swim.

gills

As they grow, they develop lungs and legs for life on land. Many lose their tails.

Groups of Amphibians

newt

Salamanders, newts, and mudpuppies keep their tails as adults.

Many have colorful skin that tells predators they are **poisonous**. Some can shed their tails to escape danger. New tails grow back later.

salamander

earthworm

Caecilians look like giant worms. They use **tentacles** to move around and find prey.

caecilian

Caecilians grab prey with their sharp teeth. They swallow their prey whole!

Frogs and toads are very similar. However, toads have shorter legs and bumpier skin.

toad

frog

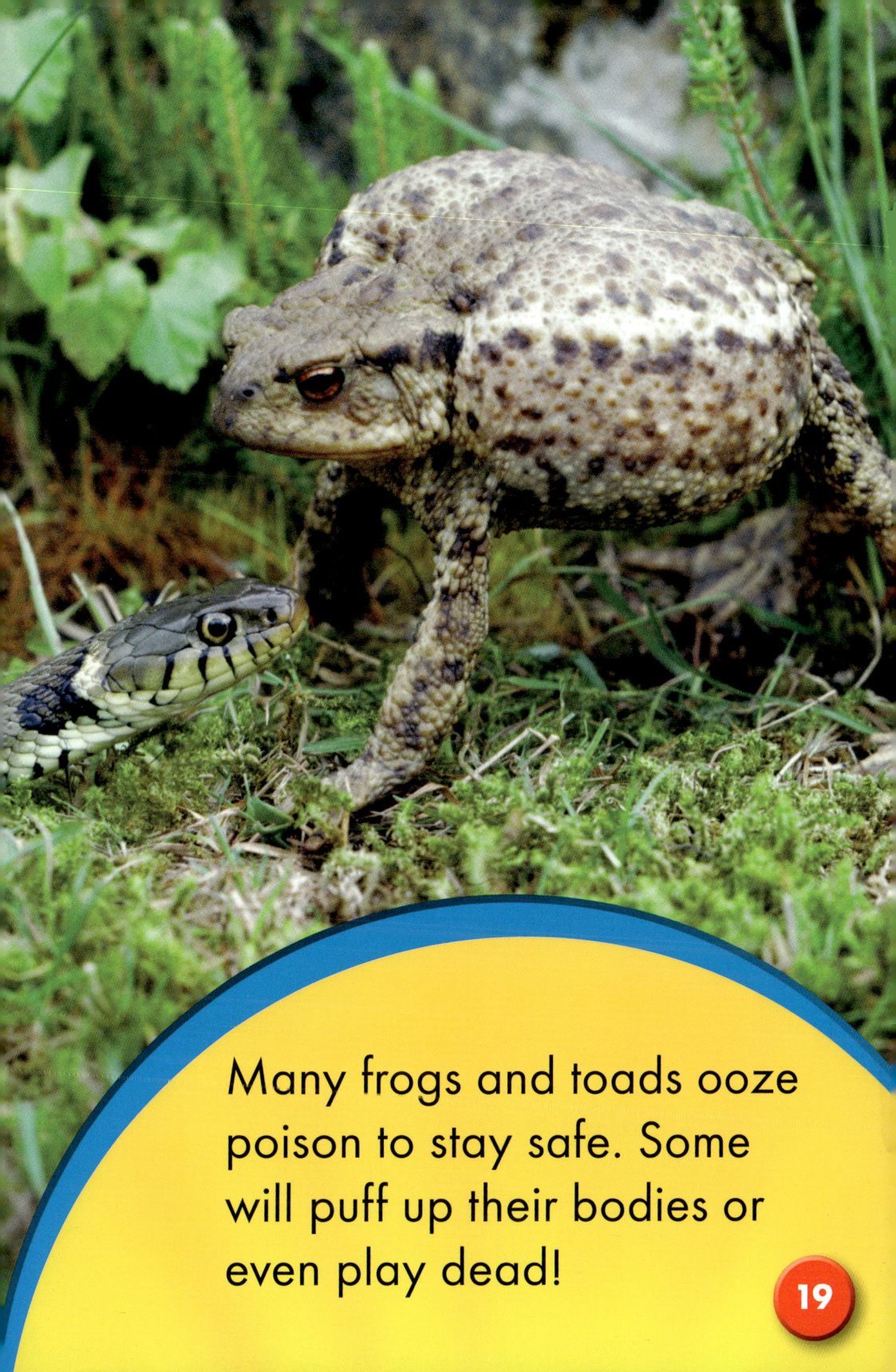

Many frogs and toads ooze poison to stay safe. Some will puff up their bodies or even play dead!

19

All-Star Amphibians

Largest:
Chinese giant salamander

Smallest:
Mount Iberia frog
and
Brazilian gold frog

Loudest:
Coquí frog

Most Poisonous:
golden poison dart frog

Lays Most Eggs:
marine toad

Longest Life Span:
olm salamander

**olm
salamander**

golden
poison dart
frog

21

Glossary

burrows—holes or tunnels in the ground that some animals dig

class—a group within the animal kingdom; members of a specific class share many of the same characteristics.

cold-blooded—having a body temperature that matches the temperature of its surroundings

extinct—no longer existing on Earth

gills—organs that allow some amphibians to breathe underwater

nutrients—elements that plants and animals need to live and grow

oxygen—a gas that animals need to live

poisonous—able to kill or harm with a poison; some amphibians ooze poison when in danger.

shed—to let fall off

tentacles—flexible extensions of the body that are used to feel and grasp

vertebrates—members of the animal kingdom that have backbones

To Learn More

AT THE LIBRARY
Berger, Melvin and Gilda. *True or False: Amphibians*. New York, N.Y.: Scholastic, 2011.

McNab, Chris. *Frogs, Toads, & Salamanders.* Milwaukee, Wis.: Gareth Stevens Pub., 2006.

Stille, Darlene Ruth. *The Life Cycle of Amphibians.* Chicago, Ill.: Heinemann Library, 2012.

ON THE WEB
Learning more about amphibians is as easy as 1, 2, 3.

1. Go to www.factsurfer.com.

2. Enter "amphibians" into the search box.

3. Click the "Surf" button and you will see a list of related Web sites.

With factsurfer.com, finding more information is just a click away.

Index